THE

RAY SOCIETY

INSTITUTED MDCCCXLIV

LONDON
MDCCCXLVI.

A

MONOGRAPH

OF THE

[TISH NUDIBRANCHIATE MOLLUSCA:

WITH

Figures of all the Species.

BY

JOSHUA ALDER AND ALBANY HANCOCK.

PART II.

LONDON:

PRINTED FOR THE RAY SOCIETY.

MDCCCXLVI.

PRINTED BY C. AND J. ADLARD, BARTHOLOMEW CLOSE.

PRINTED BY C. AND J. ADLARD, BARTHOLOMEW CLOSE.

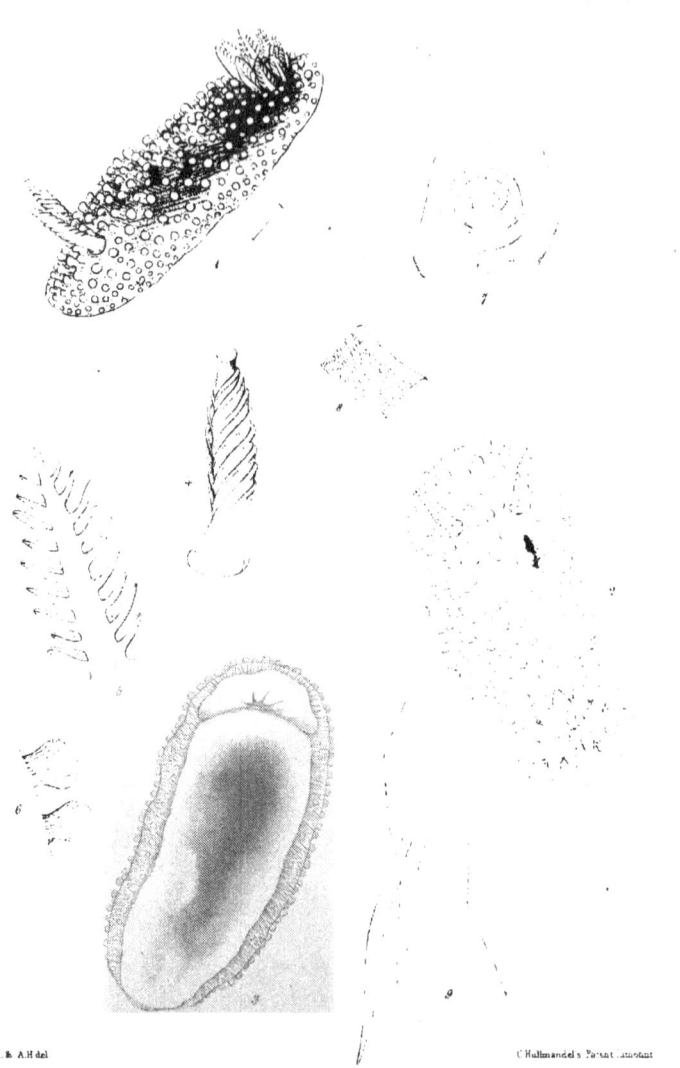

DORIS DIAPHANA.

Fam. 1, Plate 10.

DORIS DIAPHANA, ALDER and HANCOCK.

D. alba, pellucida: pallio tuberculis clavatis: branchiis 11, pinnatis, anum non propinque cingentibus.

Doris diaphana, Ald. and Hanc. in Ann. Nat. Hist. v. 16, p. 313.

Hab. Under stones near low-water mark, Meadfoot Sands, Torbay.

Body nearly half an inch long, and about half as broad, elliptic oblong, equally rounded at both ends, and not much depressed, of a pale yellowish white, very transparent, showing the intestines through the skin. *Cloak* extending a little beyond the foot all round, but rather more at the sides, covered with largish clavate tubercles, mostly of equal size, but having a few smaller ones amongst them on the back; those round the margin more thickly set and a good deal smaller. The spicula are seen through the transparent skin lying across the back, and sloping diagonally down the sides; those of the tubercles are small, and diverge towards the top. *Tentacles* nearly linear, pale yellow or yellowish white, transparent below, laminated with eight plates sloping very obliquely down behind; margins of the apertures nearly smooth, without sheaths. *Branchial plumes* eleven, simply pinnate, and partially retractile within a slight groove, set in an incomplete circle round the vent, leaving a tuberculated area within. *Head* with a large semicircular veil. *Foot* yellowish white, rounded in front and obtusely pointed behind, scarcely produced beyond the cloak, very transparent, showing the liver through the centre in a large, very dark, blackish-brown patch.

The spicula are unbranched, smooth and crystalline, a little bent in the centre, and tapering at the ends.

This *Doris* approaches nearest to *D. bilamellata*, but, besides the difference in colour and transparency, the branchial plumes are fewer, and arranged in a more circular form.

Two specimens were found under the same stone, in the locality above mentioned, about the middle of May, 1845, at which time they were spawning. The spawn is attached to stones, and forms a coil of two volutions of moderate width, sloping a little inwards at the upper margin.

Fig. 1, 2, 3. *Doris diaphana* in different positions.
    4. Tentacle.
    5. Branchial plume.
    6. Tubercles of the cloak.
    7. Spawn.
    8. A portion of the same more highly magnified.
    9. Spicula.

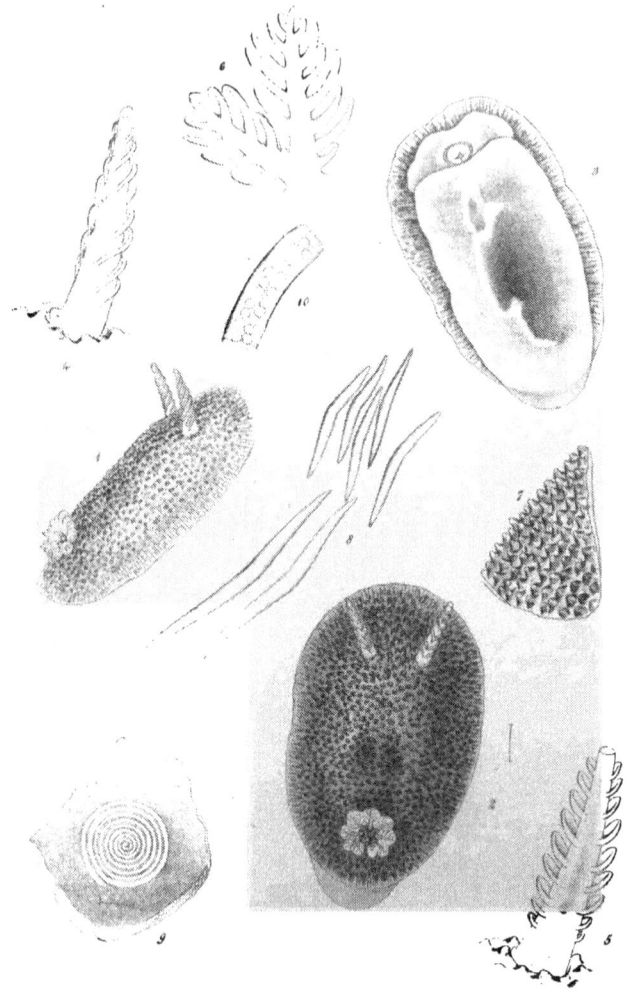

A Hancock del

C Hullmandel's Patent Lithotint

DORIS PUSILLA.

Fam. 1, Plate 13.

DORIS PUSILLA, Alder and Hancock.

D. sub-ovata, depressa; pallio lutescenti, maculis brunneis confertis; tuberculis conicis: tentaculis niveis gracilibus: branchiis 9, pinnatis, niveis, anum haud propinque cingentibus.

Doris pusilla, Ald. and Hanc. in Ann. Nat. Hist. v. 16, p. 813.

Hab. Under stones at low-water mark, Torquay.

Body about three lines long and nearly two broad, slightly ovate, much depressed. *Cloak* of a pale yellowish colour, closely freckled with dark brown, and covered with short, conical, bluntly-pointed tubercles, which are not spiculose: large spicula lie imbedded transversely across the back and diagonally at the sides, but in consequence of the dark markings they are not very conspicuous through the skin, excepting towards the margins. *Tentacles* pure white, rather long and slender, slightly tapering at the top and truncated. They have about nine laminæ, and are without sheaths. *Branchial plumes* nine, short, stout, and simply pinnate, rather obtuse at the top, of a beautiful snowy whiteness, set in a small circle, interrupted behind, and leaving a tuberculated space within. The two posterior plumes appear to arise from those next them. *Head* with a broad veil, expanded and slightly bilobed. *Foot* almost colourless, having a tinge of yellow, rather sinuated in front, and extending a little beyond the cloak behind when the animal is moving: the liver appears distinctly through it, of a chocolate brown colour.

Of this interesting and pretty little species, two individuals were found among the rocks at the Gentlemen's Cove, Torquay, during a low spring tide. From its diminutive size and flattened form it had nearly been passed over as a small *Planaria*, but a second glance showed its true character. It comes very near to *Doris depressa*, but may readily be distinguished from that species by its darker colour, and the beautiful snow-white tentacles and branchial plumes: the latter are rather larger than in *D. depressa*, and form a smaller circle round the vent. The papillæ of the cloak, too, are much shorter and stouter than in that species. The spawn of the two species is very similar, and unlike that of the other *Dorides*, forms a very thin spiral thread of many volutions, which might easily be mistaken for the spawn of an *Eolis*. In *D. pusilla* the number of volutions is nine, very closely set, and having only two or three ova in breadth throughout. It is deposited about the latter end of May.

The spicula are smooth, rather bent in the centre, and tapering to a point at each end. The diagonal ones at the sides are very large, in proportion to the size of the animal.

Fig. 1, 2, 3. *Doris pusilla* in different positions.
    4. A tentacle in profile.
    5. The same, front view.
    6. Branchial plumes.
    7. A portion of the cloak.
    8. Spicula.
    9. Spawn.
   10. A portion of the same more highly magnified.

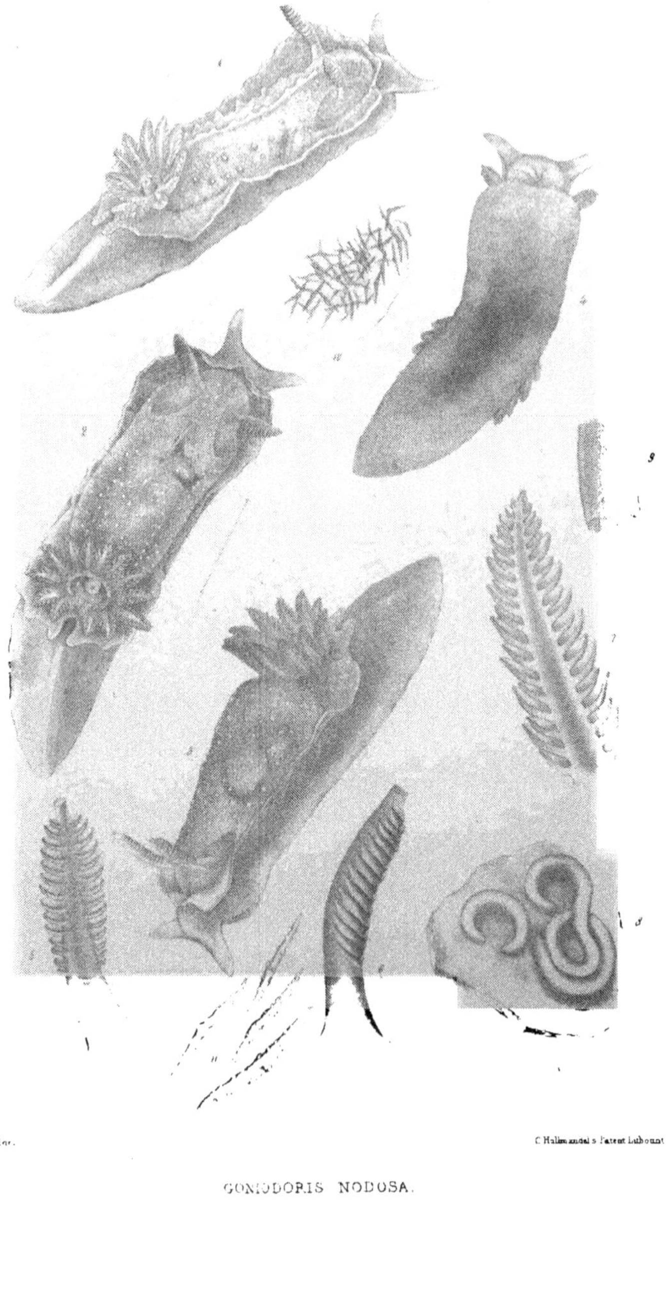

C Hullmandel's Patent Lithotint

GONIODORIS NODOSA.

Fam. 1, Plate 18.

GONIODORIS NODOSA, Montagu, sp.

G. alba, vel flavescens, carneo-tincta, maculis albis opacis: pallio, margine inciso, medio carinato, tuberculis utrinque instructo; branchiis 13, pinnatis.

Doris nodosa, Mont. in Linn. Trans. v. 9, p. 107, pl. 7, f. 2.   Flem. Brit. Anim. 282.

Doris Barvicensis, Johns. in Ann. Nat. Hist. v. 1, p. 55, pl. 2, f. 11—13.   Thomp. in Ann. Nat. Hist. v. 5, p. 87.

Goniodoris nodosa, Forbes in Ann. Nat. Hist. v. 5, p. 105.

Goniodoris emarginata? Forbes in Ann. Nat. Hist. p. 105, pl. 2, f. 12.

Goniodoris elongata? Thomp. in Ann. Nat. Hist. p. 88, pl. 2, f. 7.

Hab. Under stones and in crevices of rocks within tide marks, and in shallow water; not uncommon. Coast of Devonshire, *Montagu*.   Cornwall, *R. Q. Couch, Esq.*, Torbay.   St. Andrews, *Rev. Dr. Fleming.*   Berwick Bay, *Dr. Johnston.*   Shetland and Isle of Man, *Professor E. Forbes.* Penmaen-rhos and Llandillo Bay, North Wales, *J. Price, Esq.* Lambay Island (*G. elongata,*) Killery Bay, county Galway, and North of Ireland, *W. Thompson, Esq.*   Courtmacsherry Bay, county Cork, *Professor Allman.*   Dublin Bay and Malahide.   Tynemouth, Cullercoats, and Whitley.

Body about an inch long, ovate-oblong, sub-prismatic, rounded above, produced and obtusely-pointed behind, of a transparent white, tinged with rose- or flesh-colour, and sprinkled with very minute opake white or yellow spots.   *Cloak* subquadrangular, or somewhat fiddle-shaped, not much produced, rounded on the back, with a free margin which is scalloped into points, deeply notched behind, and generally turned upwards.   It is sprinkled, especially towards the sides and at the posterior end, with very minute opake white or sulphur-yellow spots; larger and more distant spots are disposed over the back. An elevated ridge, more or less distinct, runs down the centre, and two or more irregular rows of tubercular elevations down the sides of the back, each tipped with white.   These are readily seen in young individuals, forming generally two indistinct rows: in full-grown specimens they are more numerous and less regular, and not always distinctly visible. *Dorsal tentacles* yellowish, strongly laminated on the upper portion, with about thirteen or fourteen plates, above which the shaft is produced and truncated; the lower portion for about one third the length is smooth.   The parts of the cloak around the tentacles are slightly depressed, but appear more so from being very transparent, and without opake spots.   *Head* appearing beyond the cloak in front, having the sides produced into flat, angular, blunt, tentacular processes, occasionally tinged with yellow.   *Branchial plumes* thirteen, large, pointed, transparent and colourless; thick in the central stem, and simply pinnate, with alternate larger and smaller laminæ: they form a nearly complete circle round the vent, which is tubular.   In large individuals there are three small additional plumes which fill up the posterior opening of the circle.   Behind these there is a transparent spot, caused by a deep indenture of the cloak, which has been taken for a pore: there is, however, no perforation.   The body is extended considerably behind the cloak into a blunt-pointed tail, down the central ridge of which is a sulphur-yellow streak. *Foot* oblong, rounded in front and pointed behind, of a bright rosy brown in the centre from the liver appearing through.

Calcareous spicula are imbedded in almost every part of the skin of this animal, but nowhere crowded except in the oral tentacles and at the tail, where they form slightly radiating, dense tufts : a few pass transversely over the back and across the foot; they are more numerous in the margin of the cloak and along the sides, forming two longitudinal rows on each side, uniting at the tail. In the dorsal tentacles they are small and curiously bent to suit the cylindrical form of these organs, in which they are placed crosswise. The spicula of the body are rather large, crystalline, pointed at the ends and slightly bent in the middle, and are covered with distant irregular nodulous rings.

The pulsations of the heart are 72 in a minute.

This species appears to be common on all parts of our coast. It is liable to some variation, especially in the tubercles, which has given rise to several spurious species; at least such we are now inclined to consider them after a long and tolerably intimate acquaintance with the several varieties.

It may, however, be necessary to enter a little more into detail in justification of this opinion, in which we unfortunately differ from some of our brother naturalists. *Goniodoris nodosa* is to be found between tide-marks on the Northumberland coast nearly the whole year round, but is most plentiful in the months of April and May, at which time it is in the height of the spawning season. Large full-grown specimens may then be found in the crevices of the rocks and under large flat stones, congregated together in considerable numbers; so many as forty or fifty specimens have been found under the same stone. At this season the tints of colour are rather brighter than usual, particularly the blue patch on the right side, which is produced by that curious organ, called the purple bag by Swammerdam, appearing through the transparent skin. They are at this time generally distended with spawn, which stretches the skin so as to entirely obliterate the tubercular elevations on its surface. In this state we take it to be the *Doris Barvicensis* of Dr. Johnston. It is not uncommon, however, to find large specimens at this season with the tubercles large and conspicuous. At the commencement of the breeding season the tubercles are in all stages of development, from the most perfect down to entire obliteration, when their white apices alone mark their situation. After spawning most of the large individuals disappear, and the species is scarce on the coast for a while until the young animals begin to make their appearance, which they do in August. These are sometimes without tubercles in their earliest stage, and the skin so thin as to allow the liver to impart a pink or brownish colour to the back. This state exactly resembles the *G. emarginata* of Forbes. They soon assume the form of *Doris nodosa*, Mont., which they retain through the remainder of the season, though they continue to grow for some months afterwards. We have some suspicion that the *Doris marginata* of Montagu may be another variety of this species, but if so the figure is incorrect, as the foot is not seen beyond the cloak. *Goniodoris elongata*, Thomp., we also place, not without hesitation, as another variety : one reason for doing so is, that we collected many specimens of *G. nodosa* at Malahide (a locality near to that where *G. elongata* was obtained,) which assumed, especially when contracted in spirits, much of the elongated form of the latter animal.

GONIODORIS NODOSA.

The spawn forms a thick narrow belt, curved into an imperfect circle, and is generally of a rosy hue. The eggs are small and numerous, nearly filling the gelatinous envelope. Sometimes the coil takes a second irregular turn, having the appearance of two masses united, as represented in the drawing.

Fig. 1. *Goniodoris nodosa*, tuberculated state.

2, 3, 4. Different views of the same with the tubercles obliterated.

5, 6. Dorsal tentacles.

7. A branchial plume highly magnified.

8. Spawn.

9. A portion of the same more highly magnified.

10. A portion of the skin with imbedded spicula.

11. A few of the spicula more highly magnified.

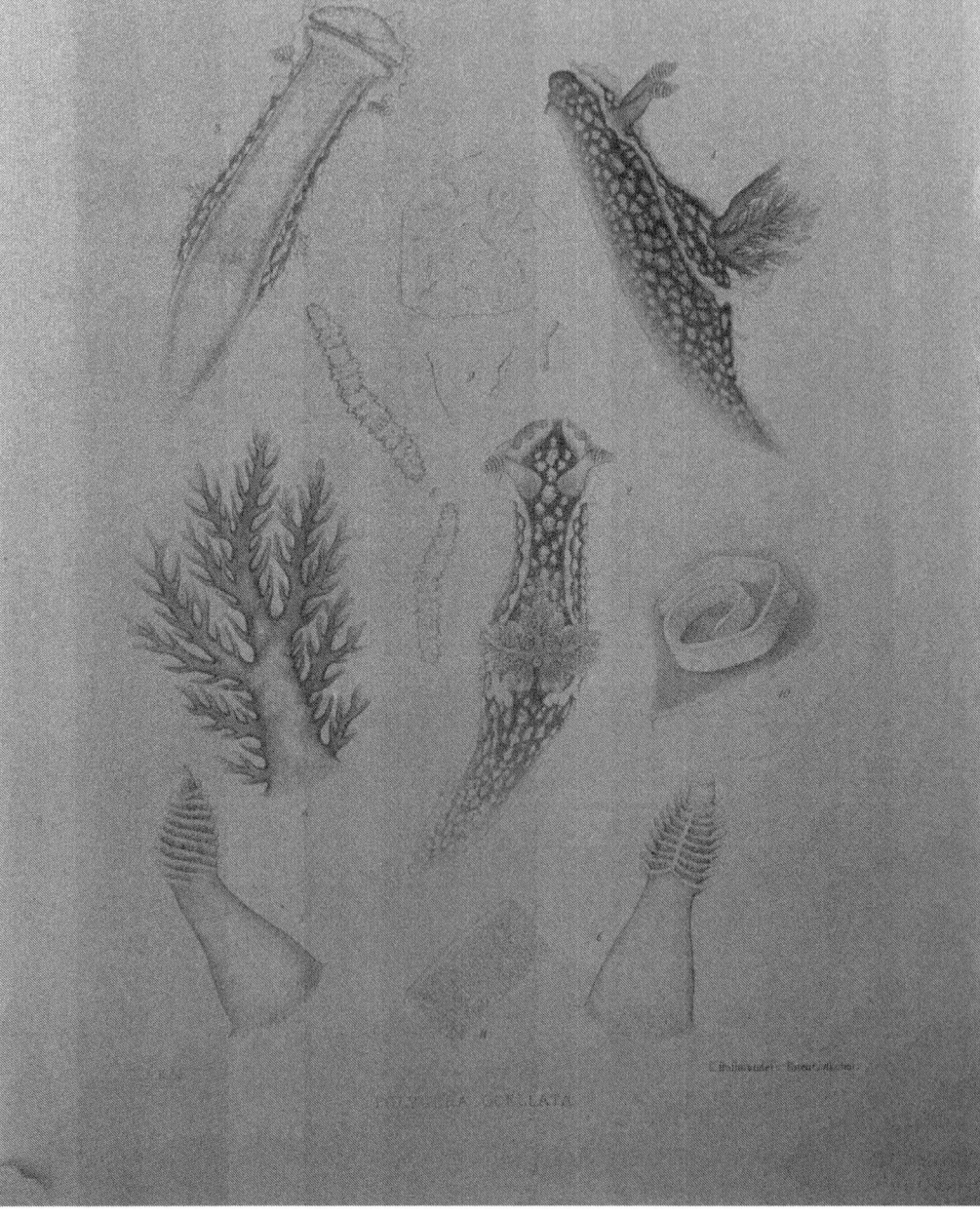

[Tab.] 31.33

HOLOTHURIA OCELLATA

POLYCERA OCELLATA, ALDER and HANCOCK.

P. viridi-nigricans, maculis magnis, tubercularibus, luteo-albidis : tentaculis elongatis, ad basin tumidis, laminis 7-8 : velo parvo, multi-lobato : appendicibus branchiarum lobato-ramosis, albidis.

Polycera ocellata, Ald. and Hanc. in Ann. Nat. Hist. v. 9, p. 33.

Triopa Nothus ? Johns. in Ann. Nat. Hist. v. 1, p. 124.

Hab. Under stones in pools between tide-marks. Cullercoats and Whitley. Torbay. Dredged in Dublin Bay.

Body varying from three eighths to three quarters of an inch in length, linear-oblong, a little contracted behind the head and tapering to a point posteriorly ; of a greenish black, arising from a very dense, minute black freckling on a yellowish ground, and covered with large spots of a pale yellow or reddish fawn-colour, the latter from the viscera appearing through. These spots are tubercular, wide at the base and sharp pointed : they are capable of extension and depression. *Dorsal tentacles* rather long, broad and conical at the base, and smooth for about two thirds up, above which they are laminated with seven or eight, sometimes nine, plates ; the apex considerably produced and truncated ; the whole is of the same colour as the body, with the sides of the laminæ darker, the edges and tip yellowish. The eyes, which are not easily distinguished on account of the dark colour of the skin, are placed a little way behind these organs. *Head* rounded in front, the sides of the mouth fleshy, slightly produced and obtusely angular. *Veil* short, interrupted in front, and continued round the sides of the head, with numerous yellowish white tubercular points ; these are extended along the sides of the body in a tuberculated ridge, contracting towards the centre of the back and expanding again near the branchiæ, which they almost surround, terminating on each side with a tuberculated, sub-ramose branchial lobe. *Branchial plumes* five, rather large and imperfectly tripinnate : the three anterior plumes nearly of equal size, the posterior much shorter, arising from the base of the lateral ones ; their colour is the same as that of the body, but not so dark ; the margins pale and sometimes stained with fawn-colour. *Foot* yellowish, freckled with obscure green and stained with fawn-colour in the centre from the liver appearing through ; truncated and rather squared in front, with the sides slightly produced ; rather abruptly tapering to a point behind.

This species very closely resembles *P. Lessonii*, nevertheless, after many careful observations of both in a living state, we feel convinced of their distinctness. The tubercles are larger, fewer, and less permanent in form, but when produced they are more pointed than in *P. Lessonii* ; those that surround the dorsal area are also generally whitish, while in the latter species they are of a deep yellow and more obtuse. But the principal difference is in the tentacles which in *P. ocellata* are longer, thicker, and more conical in the shaft, have fewer laminæ, and are more produced at the apex than in *P. Lessonii*. The branchial plumes too are larger and the branchial lobes more slender and less numerous.

These, with the difference in colour, constitute the chief distinctive characters, but there is also observable some little difference in their habitats. On the Northumberland coast, where it is not very rare, we have never found this species but within tide-marks; *P. Lessonii*, on the contrary, is generally found in deep water on the same coast, and is common upon corallines brought in by the fishing boats; but never, among the many specimens which we have obtained by this means, have we found a single individual of *P. ocellata*.

At Torquay we found this species in pools among the rocks near high-water mark.

A curious light-coloured variety was brought to us among the produce of a day's dredging in Dublin Bay, where it appeared to occur in tolerable abundance: we did not learn at what depth, but from the other contents of the dredge we conclude that it was in shallow water. This variety had the dark portion of the surface much reduced in size, and forming little more than a network round the spots. Another variety occurs on the Northumberland coast; it is of a pale yellowish green entirely devoid of spots, and might easily be confounded with the young of *P. Lessonii*, but for the tentacles and branchial plumes, which are always sufficient to distinguish the species.

The *Triopa Nothus* of Dr. Johnston is most likely the young of this species; but, from the imperfect state of the specimen from which it was described, it would be impossible to speak with certainty.

This animal floats frequently, and can fix itself by its tail with great firmness to any substance over which it may be passing. It is sometimes difficult to detach it from the polished surface of glass. Its heart beats from seventy-two to eighty-eight times in a minute. Calcareous spicula are distributed through the skin, but are not crowded nor placed in regular order. There are two kinds, one much larger than the other: the larger is cylindrical, obtuse at the ends, a little bent in the centre and pretty regularly covered with irregular circles of tubercular points; the other kind is scarcely one third the size, smooth, cylindrical, slightly bent in the centre and thickened and rounded at the ends.

The spawn occurs in July and August—occasionally in September; and is generally of a delicate rose colour and of a strap form, attached by its edge to stones: it is usually two or three times coiled, sometimes not so much, with the coils rather wide apart and angulated. The eggs are of an oval form, and mostly in transverse rows.

Fig. 1, 2, 3. *Polycera ocellata*, different views.
4. A branchial plume, very highly magnified.
5, 6. Side and front views of a tentacle, much magnified.
7. A portion of the skin, showing imbedded spicula.
8, 9. Spicula, more highly magnified.
10. Spawn.
11. A portion of the same more highly magnified.

Fam. 3, Plate 1.

LARVÆ of the EOLIDIDÆ.

The development of the ova in the *Eolididæ* is so entirely similar to what takes place in *Doris*, that we do not think it necessary here to enter into the details which will be found under the head of that genus. The coil of spawn in this family is generally more slender, and the number of volutions greater than in the *Dorididæ*; but in these respects there is considerable variation in the different groups, as will be seen on referring to the plates of species.

The larvæ undergo only a very slight variation of form throughout the entire order. For the purpose of showing the limits of this variation, we have here given a representation of some of the most distinct forms in the family *Eolididæ*, from which it will be seen how very much they resemble each other. On account of their minuteness the natural size could not be given; it must therefore be borne in mind that the largest of them are only just visible to the naked eye.

Genus 11. DENDRONOTUS,* Alder and Hancock.

Corpus elongatum, lateribus compressis, sæpissime lato altus: pallio nullo. *Caput* subinferius, velo parvo, ramoso, obtectum: maxillis corneis. *Tentacula* duo dorsalia, clavata et laminata, intra vaginas retractilia. *Branchiæ* cylindricæ, ramosæ, lineâ unicâ utrinque dorsi dispositæ. *Pes* linearis, gracilis. Orificia generationis et ani ad latus dextrum.

We have found it necessary to institute this genus for the well known *Tritonia arborescens* of authors and its allies, which are distinguished from the true *Tritoniæ* (*T. Hombergii*, &c.) by the form of their tentacles, the want of a sub-pallial ridge on the sides of the back, and the free arborescent nature of their branchiæ. These characters alone had induced us to consider them generically distinct before we had an opportunity of examining their internal structure; but it was not until we had dissected each that we were aware of the full extent of their differences; the gastric system of *Dendronotus* presenting the ramifications so peculiar to the family *Eolididæ;* while the stomach of *Tritonia* consists only of a simple pouch without appendages, as in the *Dorididæ;* the liver in the former instance being much broken up and occupying the sides of the ramifications, whilst in the latter it forms a single mass in the posterior part of the body. This important character obliges us not only to form of the species so differing a new genus, but to remove them into the family *Eolididæ*, to which, from their structure, they properly belong. Even in an anatomical point of view, however, the genus *Dendronotus* will be found to have only partially assumed the characters of this family, retaining still much resemblance to the *Tritoniadæ*, and constituting one of those transition forms so often found uniting the different types of organization in the animal kingdom.

The body is elongated, much compressed at the sides, and tapering to a point behind. The back is rounded, and has no vestige of a carinated ridge at the sides, or rudimentary cloak, as is observable in *Tritonia*. The veil is short and more or less branched, covering the head, which is sub-inferior and indistinct. There are two tentacles, which are dorsal, clavate, and transversely laminated on the upper part as in *Doris:* they are retractile within sheaths, generally branched at the margins. The branchiæ are arranged in a single line on each side of the back, rising gradually from it at their base, and dividing into cylindrical branches much resembling a tree without leaves, or, perhaps more closely, a branch of coral. This structure is different from that of *Tritonia*, whose branchiæ are composed of flattened leaflets or plumes like those of the *Dorididæ*. We have not been able to detect the ciliary movements on these organs, though we have found cilia in vigorous action along the margin of the foot and on the anal nipple: it is therefore likely that they exist over the whole of the body; and that they cover both the branchial tufts and branches of

---

* From δενδρον, a tree, and νῶτος, the back.

the veil, for we have seen the blood passing to and fro in each of these organs so as to leave little doubt of their branchial nature. The foot is linear and slender, formed as much for clasping corallines and sea-weeds as for crawling on a flat surface, though from the thin and pliant nature of its sides it may be used for either purpose. The sexual aperture is situated on the right side below the first branchial tuft, and that of the anus further behind and a little above, between the first and second branchiæ.

The digestive system varies in some respects from that of *Eolis*. The mouth and jaws are the same, the latter varying only a little in form (f. 5). The tongue occupies in the same manner the ridge of a wedge-shaped muscle that rises in the centre of the mouth, and has a similar complicated muscular apparatus for its movement. In *D. arborescens* this organ is composed of upwards of twenty transverse rows of curved, denticulated spines, with a large central one, also denticulated. Each row contains twenty of the small spines, which are brilliantly crystalline, the whole forming a very beautiful object for the microscope (f. 6, 7, 8). On account of the smallness of our specimens we were not able to detect the salivary glands, but as we did not find them outside the buccal mass, they are probably concealed by the jaws, as in *Eolis*. The œsophagus is much larger than usual, and opens into a well-defined stomachal pouch, which terminates in a short intestinal canal that opens on the right side between the first and second branchial tufts. The hepatic organ, however, shows the widest deviation from the structure of *Eolis*. The central vessel in *Dendronotus* is not a mere canal passing from the stomach and receiving the branches from the glands of the papillæ or branchial tufts, but is a large folliculated mass (f. 2, 3) occupying the centre of the body—occupying, in fact, the very position of the liver in the *Dorididæ* and *Tritoniadæ*, and communicating with the stomach by a constricted duct. From this mass branches pass off into the branchiæ and tentacular sheaths : these branches lose their follicular structure and become mere tubes as they pass into the smaller ramifications of these organs. They are lined, however, through their entire length, with the granular substance observed in the other portions of the hepatic apparatus. The sides and upper anterior portion of the stomach are covered with follicular masses, resembling in every respect the great central trunk, which, as well as the stomach, is lined with vibratile cilia. The central trunk of the digestive system lies above the ovarium, and not below it as in *Eolis*. In this respect, as well as in the glandular structure of the central trunk or mass, and in its separation from the stomach by a constricted duct, *Dendronotus* shows a deviation from the type of the *Eolididæ*, and an approximation to the *Dorididæ* and *Tritoniadæ*, thus supplying a connecting link between these two forms of gastric structure in the *Nudibranchiata*.

The vascular system is furnished with a well-developed heart, consisting of a ventricle and auricle, and in other respects does not appear to differ from the rest of the family.

The nervous system is very similar to that of *Eolis*. The cerebral ganglions are four in number (f. 9,) and are placed symmetrically, giving off nerves much in the order observed in that genus. Of these we have been able to determine ten pairs. The ganglions of the dorsal tentacles (the olfactory ganglions) are larger than usual (9 *a, a,*) and are placed at the base of the laminated portion of these organs, consequently at a considerable distance from the cerebral ganglions, to which they are united by large nerves : these form the first pair. The eyes are very small. They are composed of a well-formed pigment cup, a lens, and a cornea, enveloped in a general capsule (f. 10.) The optic nerves are very long; the eyes are therefore removed to some distance from the large ganglions. The auditory capsules contain numerous otolites of an elliptic form (f. 11, 12.)

The generative organs do not appear to differ from those of *Eolis*, with the exception that the ovarium, as before stated, lies above the central mass of the digestive system.

EXPLANATION OF THE PLATE.

Fig. 1. General view of the viscera of *Dendronotus arborescens* :—*a*, buccal mass ; *b*, œsophagus ; *c*, intestinal canal ; *d*, anus ; *e, e, e*, branches of the central mass passing into the branchiæ ; *f, f*, branches of the same passing into the tentacular sheaths ; *g*, ventricle of the heart ; *g'*, auricle ; *h*, cerebral ganglions ; *h'*, olfactory ganglions ; *i, i*, portion of the generative organs ; *j*, ovarium.

2. Digestive system :—*a*, buccal mass ; *b*, œsophagus ; *c*, stomach ; *d*, intestine ; *e*, duct connecting the alimentary canal with the liver ; *f, f, f*, branches of the central mass passing into the branchiæ ; *g, g*, branches from the same passing into the tentacular sheaths ; *h, h*, central portion of the liver.

3. A portion of the digestive system seen in profile, showing the duct connecting the liver with the stomach. The letters of this figure correspond to those of fig. 2.

4. Buccal mass.

5. Jaws deprived of their muscles.

6. Tongue.

7. A portion of the same more highly magnified.

8. Dorsal view of the same.

9. Cerebral ganglions showing the origin of the principal nerves :—*a*, ganglions of the dorsal tentacles (olfactory ganglions ;) *b, b*, the eyes.

10. The eye more highly magnified.

11. Auditory capsule with otolites.

12. Otolites more highly magnified.

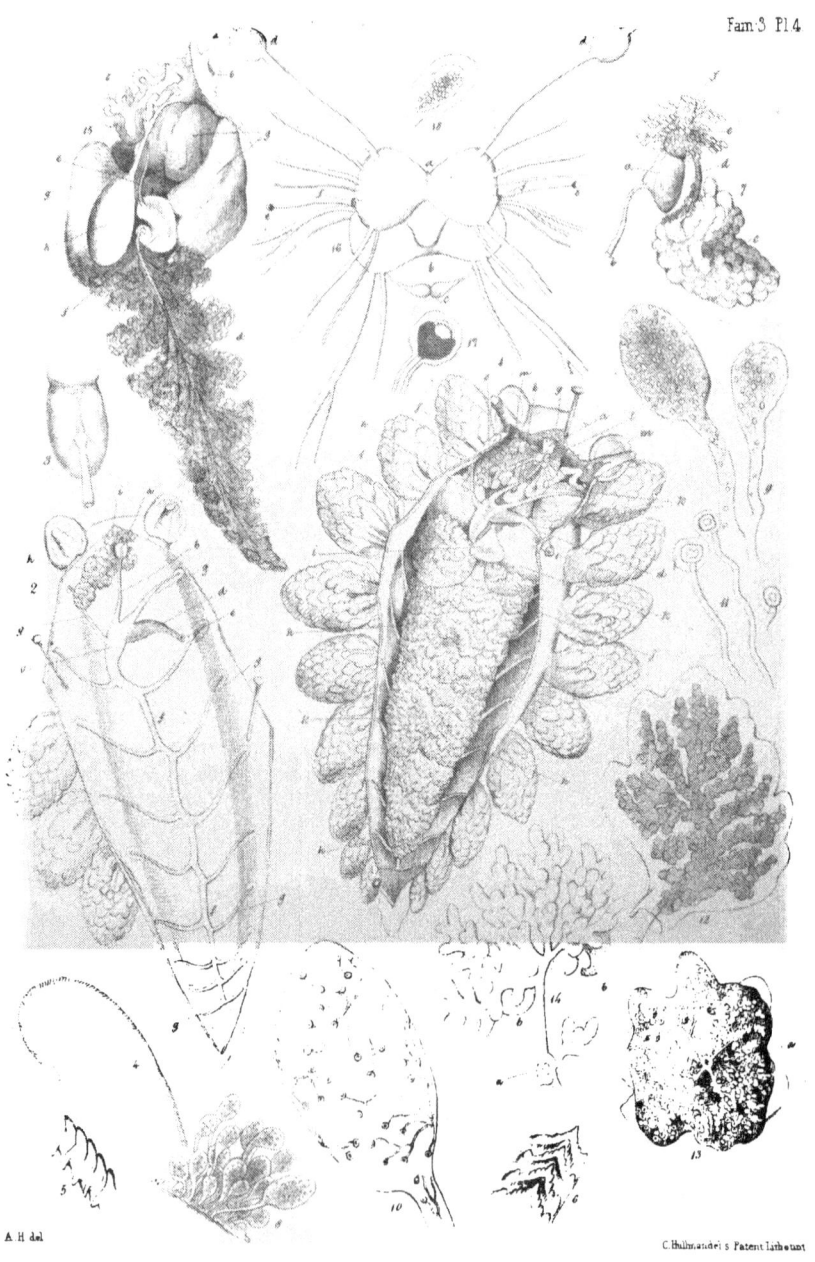

ANATOMY OF DOTO.

Fam. 3, Plate 4.

GENUS 12. DOTO,* OKEN.

Corpus limaciforme, elongatum : pallio nullo : *Caput* terminale, velo parvo obtectum. *Tentacula* duo, dorsalia, linearia, intra vaginas patentes retractilia. *Branchiæ* clavatæ vel ovatæ, tuberculis verticillatis muricatæ, lineâ unicâ utrinque dorsi dispositæ. *Pes* linearis. Orificia generationis et ani ad latus dextrum.

The genus *Doto* was established by Oken in his 'Lehrbuch der Naturgeschichte,' published at Jena in 1815,† for the *Doris pinnatifida* and *D. maculata* of Montagu. This genus has been overlooked by most of the subsequent writers on the *Nudibranchiata*, who have erroneously placed the species belonging to it in other genera—*Tritonia, Tergipes*, &c. More recently they have been referred to *Melibæa*. M. Sander Rang formed the latter genus for a mollusk obtained by him on floating sea-weed near the Cape of Good Hope, and first published it, we believe, in his 'Manuel des Mollusques,' in 1829. It approaches very near to *Doto*, but differs in having a large funnel-shaped veil fringed inside like that of *Thethys*, and a proboscidiform mouth. These differences, we think, are sufficient to warrant our considering the two genera distinct : but however that may be, the name of *Doto* has an undoubted priority, and must be adopted for our British species.

This genus has generally been considered to be allied to *Tritonia*. An examination of its internal structure, however, proves its greater affinity to *Eolis*, not only in its branched digestive system, but also in the liver being entirely placed in the branchial processes ; in which respect it shows much more decidedly the typical characters of the family than *Dendronotus*, though allied to that genus in external form, and consequently placed in the same sub-family. It is also related to *Dendronotus* by the position of the ovarium, which is above the central vessel or hepatic duct. By this circumstance, as well as in the largely developed salivary gland, it shows an alliance to the *Dorididæ*, and forms another link in the chain of affinities that unites the order.

The body of *Doto* is limaciform and elongated, rounded on the back, and without any cloak or carinated ridge. The head is covered by a short veil, plain at the edges. The mouth is small, round, and puckered. There are two tentacles which are slender and linear, retractile within sheaths, usually expanded and curved outwards at the top : they are situated dorsally nearly above the mouth.

The branchial processes are arranged in a single row along each side of the back. They are of an ovate or clavate form, slightly pedunculated, and covered with transverse rows of bluntly-pointed tubercles, capable of extension and contraction. These processes

---

* Doto, a sea-nymph.

† The name has since been used by De Haan for a genus of *Crustacea*, published in 1836 ; and we believe also, by Guerin, for another genus of crustacea, during the same year.

very readily fall off, and possibly may be detached, as Professor E. Forbes supposes, at the will of the animal, but of this we have not been able to satisfy ourselves. Their deciduous character has been thought by some to offer a decided objection to their being considered as branchiæ, and indeed the animal is quite capable of surviving the loss of the whole of these processes; but as it is now very well ascertained that the mollusca of this order respire also by the skin, even in cases where the branchial organs are more decidedly speciallized, we see no good reason for withholding the name of branchiæ from these organs, though they may perform that function less perfectly than the processes of some other genera of the order. The foot is linear, and generally rather narrow. The anus is dorsal and lateral, forming a small nipple on the right side between the first and second branchial processes. The aperture of the generative organs is placed below the first branchial process on the right side.

The anatomy of *Doto*, notwithstanding its general similarity to that of *Eolis*, differs from it in some interesting particulars.

The digestive system has the buccal mass much smaller than usual, of an oval form, and unprovided with corneous jaws. The tongue is curved and placed in the mouth as in *Eolis*; it is stiff, slender, and exceedingly small, and is composed of upwards of a hundred transverse plates each supporting a central tooth, directed backwards, and appearing, when highly magnified, a little denticulated at the sides. From the small size and weak armature of the mouth, compared with other genera, we think it likely that the animals of this genus feed only on the soft parts of the corallines, on which they are generally found. The salivary gland is ample. It lies above the generative organs on the left side of the body, and extends forward to the buccal mass, opening by a slender duct immediately before it through the inferior wall of the channel of the mouth. There is likewise another gland connected with the channel of the mouth which is probably accessory to the above, though it differs from it in structure. It is much smaller, and is composed of numerous minute oval sacs, each furnished with a delicate and suddenly constricted duct: these sacs surround the channel of the mouth, and pour the secretion into it on all sides just where it receives the duct from the large salivary gland. With a high magnifying power they appear to be covered over the inner surface with nucleated cells, having minute vessels passing from them, which ramify over the inner surface of the sacs and unite as they descend towards the ducts. The size of the vessels corresponds exactly with the diameter of the nucleus.*

The œsophagus is long, very slender, and passes from the upper posterior dorsal aspect of the buccal mass, increasing a little in size as it reaches the stomach, which is small and elongated. The interior is minutely granulated, but does not appear to be raised into folds or wrinkles. The intestine is very short and wide, especially at its junction with the stomach, which takes place further forward than usual on the right side: the inner surface is plicated longitudinally; the plicæ end abruptly as they reach the stomach. From this peculiarity of structure, and the great width of the intestine, it seems probable that it may have some other function to fulfil than that of conveying off the refuse of digestion. May it not be in this canal that the nutritive portion of the food is absorbed into the digestive

---

* We have been thus minute in the description of this gland, as we think it likely to throw some light on the ultimate structure of glands in general. We should suggest the probability that the nucleus of the cell of glandular tissue may be nothing more than the orifice of a minute vessel, or perhaps, in some cases, a mere opening through which the secretion passes.

system? The anus opens on the right side of the back, as stated above. Two delicate vessels or ducts pass from the anterior portion of the stomach to the first pair of branchial processes: the other branchial processes receive similar vessels from a central vessel which passes from the posterior extremity of the stomach, and extends the whole length of the body beneath the ovarium. The lateral vessels alternate a little, and are not always symmetrical: the last pair arise from the extreme posterior point of the central vessel.

The hepatic glands are very highly organized, and almost fill the branchial processes, with the exception of small passages hollowed out of the glandular mass, probably for the circulation of the blood. There are no distinct sheaths as in *Eolis*, though we have observed between the gland and the outer skin in *D. coronata*, a sort of network of fleshy columns similar to what we have seen between the outer and inner sheaths of the papillæ of *Eolis papillosa*. On placing one of the smaller branchial processes of *D. fragilis* in the compressor of the microscope, the gland is discovered to be of a highly racemose character: but for the complete understanding of its structure it is necessary to make a transverse section of one of the larger processes, when the gland is perceived to be highly complicated and delicately branched. The branches radiate from a minute central duct, and are much divided and crowded with follicles: their extremities reach into the papillæ. The completeness of this organ as a secreting apparatus must be evident to physiologists, as well as the improbability of its subserving any other function.

Respiration is undoubtedly performed by the surface of the body as well as by the branchial processes. Immediately beneath the skin of the body in *D. coronata*, is the same network of fleshy columns observed in the branchial processes of that species; and both in it and in *D. fragilis*, the entire surface is covered with vigorous vibratile cilia. Cilia are also on the branchial processes, but are most vigorous towards their bases.

The heart is as completely organized as in any of the order: the ventricle is strong and muscular, the auricle thin and membranous.

The nervous system is furnished with four cerebral ganglions of nearly equal size, and two small buccal ganglions of the usual form. The nerves are arranged much as in *Eolis*. There appear to be about ten principal pairs, and a few smaller ones, which we were unable to trace. The olfactory ganglions are at some distance from the cerebral ganglions to which they are united by large nerves: they are round, and placed at the bases of the tentacles. The eyes are also placed a little way from the cerebral ganglions, and are very small. The otolites of the auditory capsule are numerous and elliptical.

The generative organs do not materially vary from those of *Eolis*. The ovarium, however, is above the great central vessel of the digestive system.

EXPLANATION OF THE PLATE.

Fig. 1. General view of the viscera of *Doto fragilis*, seen from above:—*a*, buccal mass; *b*, œsophagus; *c*, stomach; *d*, intestine; *e*, anal nipple; *f*, salivary gland; *g*, accessory salivary gland; *h, h, h, h*, lateral vessels or ducts leading from the branchial glands to the central vessel connected with the stomach; *i*, ventricle of the heart; *j*, auricle of the same; *k, k, k*, generative organs; *l*, cerebral ganglions; *m, m*, olfactory ganglions.

    2. The digestive apparatus exposed:—*a*, buccal mass; *b*, œsophagus; *c*, stomach; *d*, intestine; *e*, anus; *f, f*, great central duct; *g, g, g, g*, lateral ducts from the branchial glands; *h*, salivary gland; *i*, accessory salivary gland.

DOTO.

Fig. 3. Dorsal view of the buccal mass.
    4. Side view of the tongue.
    5. A portion of the same more highly magnified.
    6. A portion of the same seen from above.
    7. Buccal mass with salivary gland :—*a*, side view of the buccal mass ; *b*, œsophagus ; *c*, salivary gland ; *d*, duct of the same ; *e*, accessory salivary gland ; *f*, channel of the mouth.
    8. A portion of the accessory salivary gland more highly magnified.
    9. Two of the sacs of the same, increased in size.
   10. A single sac of the same, still more highly magnified, exhibiting the nucleated cells and the vessels ? leading from them (the cells on the under side are not shown.)
   11. Three of the nucleated cells with the vessels attached.
   12. A lateral view of one of the smaller branchial processes, as seen in the compressor of the microscope.
   13. A transverse section of one of the larger branchial processes showing the complicated structure of the gland :—*a*, central duct.
   14. A single branch of the same more highly magnified :—*a*, central duct ; *b, b*, blood passages.
   15. Generative organs.
   16. Cerebral ganglions and nerves :—*a*, upper ganglions ; *b, b*, lateral or under ganglions ; *c*, buccal ganglions ; *d, d*, ganglions of the dorsal tentacles (olfactory ganglions ;) *e, e*, eyes ; *f, f*, auditory capsules.
   17. Eye of *D. coronata*.
   18. Auditory capsule of the same.

Fam. 3, Plate 6.

DOTO CORONATA, Gmelin, *Sp.*

D. flavescens, rubro maculata: velo truncato: branchiis utrinque 5-7; ovato-clavatis, tuber-culatis, apicibus puniceo punctatis.

Doris coronata, Gm. 1, p. 3105, No. 19.
Doris pinnatifida? Mont. in Linn. Trans. v. 7, p. 78, pl. 7, f. 2, 3.
Tritonia pinnatifida? Flem. Brit. Anim. 284.
Tritonia coronata, Lam. Anim. s. vert. 2d ed. v. 7, p. 454.
Tergipes coronata, D'Orb. in Mag. de Zool. 1837, v. 5, pl. 103.
Scyllæa punctata, Bouch. Chant. Moll. de Boul.
Melibæa coronata, Johns. in Ann. Nat. Hist. v. 1, p. 117, pl. 3, f. 5-8.
Melibæa ornata, Ald. and Hanc. in Ann. Nat. Hist. v. 9, p. 34.
Doto coronata, Lovén in Arch. Skand. Nat. p. 151.
Hab. On corallines from deep water, not uncommon; also among the rocks at low-water mark. Frith of Forth, *Dr. Grant.* Berwick Bay, *Dr. Johnston.* Isle of Man, *Professor E. Forbes.* Woodside, near Liverpool, *J. Price, Esq.* Glendore Bay, county Cork, *Professor Allman.* Dublin Bay and Malahide. Oban Bay, Argyleshire. Cullercoats, Whitley, and Newbiggin, Northumberland. Marsden, Durham. Torbay and Salcombe, Devonshire.

Body about half an inch long, nearly linear, slender, smooth, semitransparent, pale yellow or buff, spotted with reddish or brownish purple. The spots usually form a broad interrupted line from the front of the head almost to the tail; from this, belts of similar spots pass transversely between the branchial processes and blend with a line of spots that passes along each side. *Tentacles* filiform, truncated at the tip, and transparent, issuing from rather long trumpet-shaped sheaths; wide at the top and more produced anteriorly; set near together at the base and diverging above. *Veil* entire, straight in front, produced and rounded at the sides, and capable of extension and contraction in a lateral direction. *Branchiæ* from five to seven pairs; the last frequently rudimentary, the rest nearly of equal size, large in proportion to the animal, elliptical, pedunculate and muricate, semitransparent, having four or five circles of tubercles and a terminal one at the apex: the tubercles are capable of extension and contraction, and each is surmounted by a dark red spot, sometimes nearly black. The central portion of the branchiæ is generally of a brownish rose-colour, sometimes purple-brown, and occasionally pale buff or olive. *Foot* transparent white, tinged with yellow, rounded and somewhat bilobed in front; the margin slit transversely; ending in an obtuse point behind.

The heart forms a slight swelling on the back, between the first and second pairs of branchial processes. It pulsates 60 times in a minute.

This is a very beautiful and delicate little animal. It is liable to great variation in colour, and the spots are frequently jagged, and produced into streaks. The body is occasionally almost colourless, and nearly devoid of any sort of markings; and we have seen it entirely of a beautiful rose colour, with the body nearly covered with confluent purple or puce-coloured spots, relieving with great force the delicate rose-coloured branchial processes. Sometimes the body is bright yellow, with the spots both on it and on the branchiæ, of a fine carmine; and another variety occurs with the branchial processes nearly colourless, being tinged only with a watery green, and having the spots almost black.

Doto coronata is found on all parts of our coast that have been properly examined, and may be considered amongst the most common of our deep water species. Its favourite haunt is amongst the corallines that grow at a short distance from the shore, in between fifteen and twenty fathoms water: of these it appears to prefer *Plumularia falcata* and *Sertularia abietina*. Dr. Johnston states that he found about a dozen specimens on one tuft of *Plumularia Catherina*. It is also frequently found on *Sertularia pumila* among the rocks at low-water mark.

This species is found on the French, Dutch, and Swedish coasts, so that its range probably extends through the whole of the European seas.

The spawn is always found on corallines; the ova are white and deposited in a thick gelatinous riband, which is folded on the stem of the coralline in a zig-zag manner. July appears to be the height of the breeding season; spawn, however, occurs both in June and August.

After having been referred to at least six genera by different authors, this little animal, it is to be hoped, may at last find a resting-place in the genus to which, following Dr. Lovén, we have now consigned it.

The *Doris pinnatifida* of Montagu, which we place with doubt as a synonym of this species, is still involved in much obscurity. We had hoped, while on the Devonshire coast, to have satisfactorily made out that species, especially as we explored Montagu's favourite dredging grounds, and, while in Salcombe estuary, had the assistance of the boatman who was accustomed to attend him. Our search for this and some others of Montagu's lost species was, however, without success. *Doto coronata* was found at Salcombe, and we also met with the same species in Torbay as well as *Doto fragilis*. Is it not possible, therefore, that one of these may really be the *Doris pinnatifida?* Dr. Johnston thought that he recognized it in the latter, but we are of opinion, from its small size, slender form, and the spots on the tips of the branchial tubercles (an almost invariable character in *D. coronata*, but never found in *D. fragilis*,) that it is much more likely to belong to the former species. There is, in fact, nothing in Montagu's description, if we except, perhaps, the colour, and an additional row of tubercles on the branchiæ,* which does not agree with the characters of *D. coronata*. The figure is less like it, but some allowance must be made for the evident want of skill in the artist. The number of branchial processes is not mentioned in the description, but nine pairs are represented in the figure, a number beyond what we have ever found in *D. coronata*, though in one instance we have seen it with eight pairs; it is, however, still less like the young of *D. fragilis*, which, when no larger than Montagu's specimen, has only six pairs of branchial processes, and no more than three or four rows of papillæ, and is, moreover, entirely without spots. Further observations may throw additional light on the subject, but should

---

* M. D'Orbigny's *Tergipes coronata*, which we have given as a synonym of our species, has the same number of rows of tubercles as Montagu's *Doris pinnatifida*.

no other species be found agreeing better with the description and figure, the occurrence of *D. coronata* on that part of the Devonshire coast may afford a fair presumption of its identity with the long lost *Doris pinnatifida*.

Another of Montagu's species, referrible to this genus, stands in a somewhat similar predicament; namely, *Doris maculata*, which agrees with the present species in the form, colour, and markings of the body, but differs in the branchiæ being much smaller and with only one whorl of tubercles. We have occasionally found *D. coronata* with one or two of the processes of a similar form to this, either from the individuals being young, or from new branchiæ being in process of reproduction where others had fallen off. We have never met with an individual with all the branchiæ in this state, but the possibility of such an occurrence induces us to look upon this species also with some suspicion.

A more complete examination of the species on the southern coast is very desirable.

Fig. 1, 2, 3. *Doto coronata*, different views.
    4. Spawn.
    5. A portion of the same more highly magnified.

Abel

C. Hullmandel's Patent Lithotint.

EOLIS CORONATA

Fam. 3, Plate 12.

EOLIS CORONATA, Forbes.

E. gracilis, albida : branchiis oblongis, sub-linearibus, rubris, cœruleo-tinctis, apicibus albis, in fasciculis 6-7, digestis : tentaculis dorsalibus valdè annulatis, annulis 7-8 : angulis anterioribus pedis productis.

Eolida coronata, Forbes in Athenæum for 1839, No. 618, p. 647.

Hab. Under loose stones and rocks between high and low water mark. Shetland, *Professor E. Forbes.* Whitley, Cullercoats, and Newbiggin, Northumberland. Marsden, Durham. Rothesay Bay, Isle of Bute. Dublin Bay and Malahide, Ireland.

Body about an inch long, slender, nearly linear and tapering to a fine point, of a transparent watery white, tinged with rose-colour and buff, the latter from the viscera shining through. Immediately behind the dorsal tentacles there is a large longitudinal stain of rose-colour, caused by the œsophagus, and in front of them is a lozenge-shaped spot, of opake white or blue, from which a streak of the same passes to the anterior margin of the head, and another streak passes backwards between the tentacles, and terminates a short way behind them. There is likewise a white streak on the ridge of the tail, and the upper surface of the body is sprinkled over with irregular faint spots, of opake white, not unfrequently tinged with blue. *Dorsal tentacles* fawn-coloured, with a pale sulphur-yellow streak in front, subclavate, having seven or eight wide membranous rings, and incomplete intermediate ones, which gives them a peculiarly elegant appearance ; points truncated. They spread gradually apart above and approximate at the base, and are not much inclined forwards. The portion of the body bearing these tentacles is rather elevated. *Oral tentacles* much longer than the dorsal ones, tapering, white, tinged with pink or blue, and generally held in a gracefully curved position : they gradually enlarge at the base, so as to form a continuous outline with the head. Their tips have an opake white streak. *Branchiæ* elliptic-oblong, nearly linear, cylindrical, set in six or seven clusters down each side of the back. Their central vessel is of a deep crimson, varying occasionally to brick red, orange, or more rarely pale rose-red or flesh colour, with the extremities darker ; the sheaths always reflecting more or less of a bright ultramarine blue, which generally forms an oblong blotch or streak in front. The apices have an imperfect opake white ring, expanded in front and prolonged into a streak which passes a little way down the papillæ. From above, the apex appears to be perforated ; an appearance caused by the white ring being confined to the surface, and the central part transparent. The first cluster of branchiæ, which forms with the opposite one a kind of ruff round the neck, consists, in fine full-grown specimens, of twenty or thirty papillæ, set in transverse rows of five or six each, rather long on the back and diminishing towards the sides. The second cluster, divided from the first by a short space, is less numerous, and the rest become gradually less in number and size, sometimes nearly coalescing. *Foot* slender, extending beyond the branchiæ behind and tapering to a fine point, the anterior margin is slit transversely, and the upper laminæ notched in the centre ; the lateral angles produced at the sides and curved backwards.

This fine species—one of the most beautiful of its tribe—occasionally reaches an inch and a half in length. It may be distinguished by the brilliant blue which is always more or less present on the branchiæ, sometimes only giving them a faint tinge, at others forming a streak down the front of each papilla; and sometimes a very beautiful variety occurs with the blue nearly covering their whole surface. The branchiæ are variable in number and liable to fall off. An entire clump is sometimes wanting.

This is a very active animal, and whilst gliding forward its tentacles are in continual motion, bending gracefully in all directions. Both pairs appear to be endowed with great sensibility, contracting themselves and suddenly shrinking backwards on coming in contact with anything. Sometimes they will do this without apparent cause. The branchiæ, too, usually partake of this restless motion, and are capable of great extension and contraction at the will of the animal. We have had several opportunities of noticing the carnivorous propensities of this species, which is certainly not the least voracious of its tribe. After having been for a day or two without food, they will even devour their own kind, the weaker falling a sacrifice to the cravings of the stronger. Large individuals will content themselves with plucking off each other's papillæ; but should a smaller specimen be within reach, it is most mercilessly attacked, the more powerful animal laying hold of any part of the weaker that may happen to be nearest. The tail, however, is generally first seized, and fierce and determined is the onset. The devourer raises and shakes his papillæ in the manner that the porcupine shakes its quills when irritated, and then, laying back the dorsal tentacles and curling up the oral ones, fixes the protruded mouth and jaws upon his prey, when, with a convulsive shrinking up of the body, morsel after morsel is appropriated. In this manner it is not uncommon to see an individual entirely devour another, half its own size. We have also seen this species feed upon a *Lucernaria*.

Eolis coronata spawns most abundantly in June, at which period it is rather plentiful among the rocks at Whitby and Cullercoats; patches of spawn, however, are not unfrequently found in July, and occasionally in August. The spawn is attached to the underside of stones, and is disposed in a close-set spiral coil of four volutions, consisting of a waved gelatinous thread, with yellowish imbedded ova.

This species was first discovered by Professor E. Forbes in Shetland, and we have since found it in several places; principally on the northern shores of our islands. On the Northumberland coast it is one of the most common species, and in Malahide Bay we found an orange variety of it in considerable numbers on large *Laminariæ* and sponges dredged in shallow water.

We suspect that the *Eolis peregrina*, mentioned by Dr. Grant as having been found in the Frith of Forth, is this species, which somewhat resembles that of Cavolini in the colour of the branchiæ, but not in their arrangement. We see no good reason for believing that the true *E. peregrina* has ever been found in this country.

The pulsations of the heart are about sixty-five in a minute.

Fig. 1, 3. Back and foot views of *Eolis coronata*.
    2. Side view of the blue variety.
    4. Two of the papillæ more highly magnified.
5, 6. Dorsal tentacles.
    7. Spawn.
    8. A portion of the same more highly magnified.
    9. Two teeth from the tongue highly magnified.

Fam. 3, Plate 15.

EOLIS PUNCTATA, ALDER and HANCOCK.

E. carneo-lutescens, maculis albis undique aspersa: branchiis oblongis, acutis, flavido-fuscis, maculis albis, in fasciculis 5-6 digestis; tentaculis dorsalibus obliquè laminatis; tentaculis labialibus longis: angulis anterioribus pedis valdè productis.

Eolis punctata, Ald. and Hanc. in Ann. Nat. Hist. v. 16, p. 315.

Hab. In rather deep water, Torbay.

Body about an inch long, nearly linear, pellucid yellowish white, tinged with rose-colour about the head and shoulders, and of a buffish hue behind, from the viscera appearing through; the whole spotted over with rather large, distant, opake white spots. *Dorsal tentacles* slightly conical, tapering towards the top and truncated, having twelve or thirteen very oblique laminæ and imperfect intermediate ones: the laminæ slope downwards behind as in the *Dorididæ*, and are interrupted in front, exposing the shaft, as is also the case in several of that family; they are united behind to an elevated zigzag ridge. The colour of the tentacles is a dull yellow, opake and paler towards the top; they stand apart above and approach towards the base, inclining very little forwards. *Oral tentacles* twice the length of the dorsal ones, and tapering, their bases forming the sides of the head; pellucid white, slightly tinged with rose-colour, and having an opake yellowish white streak towards the tips. *Branchiæ* in six or seven clusters; the anterior pair the largest, and divided from the second pair by a considerable naked space, in which the heart is situated. The first cluster has three principal rows, containing altogether between thirty and forty papillæ; those next the foot small and closely set. The second cluster has about half as many papillæ; the remaining clusters are small and nearly confluent, approaching very near to the tail. The papillæ are elliptic oblong, tapering above, and of a dark brownish flesh-colour, spotted with white in the same manner as the body; the spots becoming smaller and clustered towards the apex, which is transparent. *Foot* with the sides nearly parallel, the posterior extremity rather abruptly tapering, and terminating a short way behind the branchiæ; the front deeply bilobed and extended into long tentacular processes, curved backwards at the sides, the outline having a bow-like appearance; the margin is transversely slit. Colour transparent ·yellowish white or buff, with a tinge of flesh-colour; the upper part spotted with white.

This is another of the novelties that we have obtained in Torbay, where it was dredged in deepish water off Berry Head. In general appearance it somewhat resembles *E. Drummondi*, but is at once distinguished from this and all its congeners by the dorsal tentacles, which are laminated in oblique folds like those of *Doris*. It is also peculiar from the conspicuous white spots that cover it on all sides.

This interesting animal lived with us for some time, and afforded an opportunity of studying its habits. It is a very active creature, moving about in all directions, and frequently swimming inverted on the surface. It would appear also to be rather unscrupulous in its voracity, as in one instance we suspect it to have made a repast of its own spawn; and on other occasions it devoured a *Doto coronata*, a portion of an *Eolis coronata*, and several individuals of *E. olivacea*. When it seized its prey the body shrunk up, the papillæ became agitated and twitched in a peculiar manner, the tentacles were thrown back, the lips retracted, and the mouth advancing, in an instant the object of attack was forcibly drawn within reach of its formidable jaws; at short intervals the attack was renewed; the animal, however, all the while keeping its hold until the whole was despatched. On the first seizure of its prey, this species generally emits a sound resembling the click of a watch, or, perhaps more nearly, the sound produced by pressing the edges of the finger and thumb nails together, and then letting them slip. The sound is generally repeated on each renewal of the attack, and is frequently produced at other times. We have heard it on several occasions when the animal was perfectly quiescent, and again when it was moving about, probably in search of food. The sound, which is never emitted more than once at a time, and generally at considerable intervals, is sufficiently loud to be heard at some little distance. From the nature of the sound we are quite inclined to agree in opinion with Dr. Grant, who was the first to notice that these creatures have the power of emitting sounds, that it is produced by the cutting edges of the corneous jaws, and indeed it would be difficult to understand by what other means it could be effected.

Whilst in confinement, this species deposited three patches of spawn at different times. The eggs are white, and are contained in a wide, undulating gelatinous cord, coiled about three times. One of the patches which was deposited on the 29th or 30th of June, was hatched on the 9th of July. The larvæ very closely resemble those of *E. coronata*.

Fig. 1, 2, 3. Different views of *Eolis punctata*.
    4. Two of the papillæ highly magnified.
5, 6, 7. Side, front, and back view of a dorsal tentacle, much increased in size.
    8. Spawn.
    9. A portion of the same more magnified.

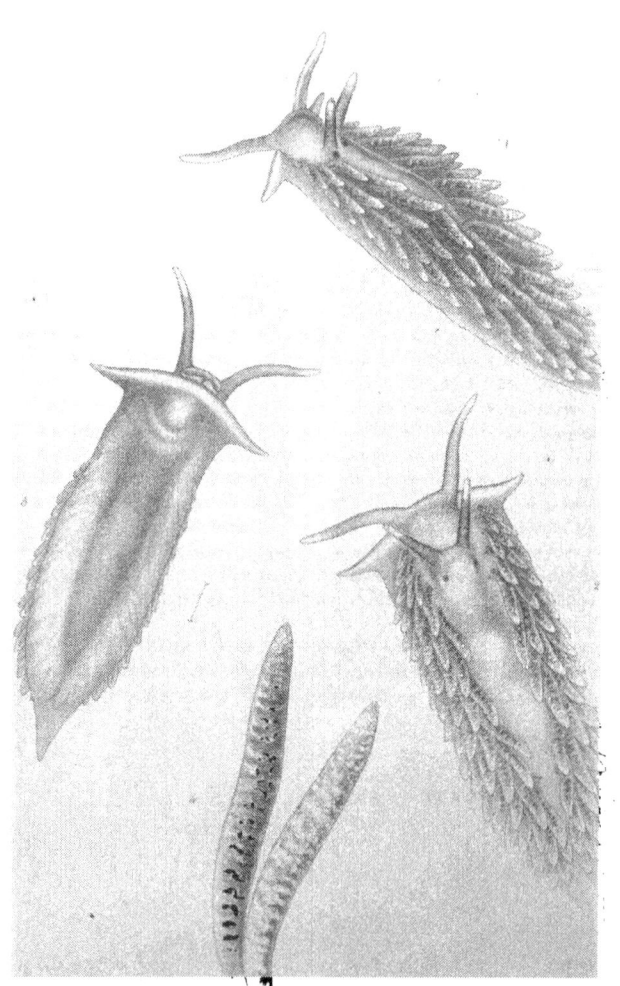

Fam. 3, Plate 23.

EOLIS ANGULATA, ALDER and HANCOCK.

E. subangulata, depressa, pallidè aurantiaca: branchiis cylindricis, aurantiacis, albo maculatis; tentaculis brevibus; angulis pedis acutis.

Eolis angulata, Ald. and Hanc. in Ann. Nat. Hist. v. 13, p. 165.

Hab. On a stone from the fishing boats, Cullercoats.

Body about four lines long, depressed, subangulated, broad in front, and terminating rather abruptly behind, of a pale pellucid orange colour. *Dorsal tentacles* short, conical, obtuse, orange tipped with white; set a little apart and only slightly inclined forwards. *Oral tentacles* rather longer than the dorsal ones, nearly linear, rather obtuse, the lower portion transparent, the rest of an opake white. The eyes are very large, and placed a little behind the dorsal tentacles. *Branchiæ* cylindrical, rather long, slightly elliptical, obtusely pointed, orange-coloured with white apices; the surface covered with opake white blotches. The central vessel is slightly granular and nearly fills the sheath, except at the top, the points extending considerably beyond it. They are arranged along the sides in ten or twelve close-set, but well-defined, rows of four or five papillæ each, leaving a broad space on the back. *Foot* broad, transparent, and delicately tinged with orange. It extends only a short way beyond the branchiæ behind, where it suddenly tapers to a fine point: it is broad in front and produced into lengthened angles at the sides.

A single specimen of this species was obtained from a stone brought in by the fishermen at Cullercoats, probably from deepish water. It lived several days, and moved about with much ease and rapidity, swimming at the surface of the water much more quickly than usual.

Fig. 1, 2, 3. *Eolis angulata*, different views.

4. Two of the papillæ more highly magnified.

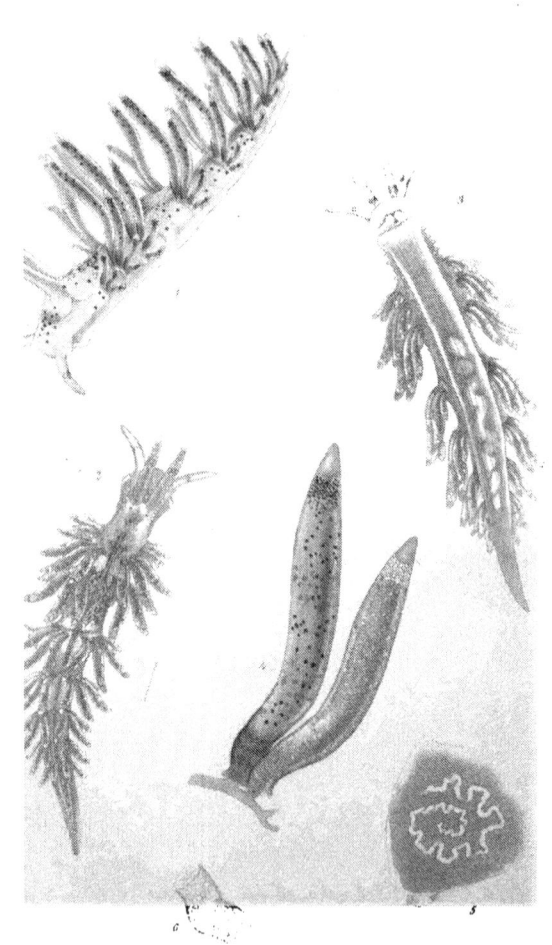

Fam. 3, Plate 30.

EOLIS AMŒNA, ALDER and HANCOCK.

E. gracilis, virescenti-alba, brunneo notata, punctis albis: branchiis viridibus, luteo-maculatis, in seriebus 8 digestis, quorum 3, contiguæ et 5 remotæ sunt: tentaculis dorsalibus longis, brunneo-cinctis: lateribus anterioribus pedis rotundatis.

Eolis amœna, Ald. and Hanc. in Ann. Nat. Hist. v. 16, p. 316.

Hab. Dredged in Torbay.

Body about three lines long, slender, nearly linear, terminating behind in a fine point; of a greenish white, tinged with yellow from the viscera appearing through; on the back are a few dark brown markings. The head and shoulders are sprinkled with opake white tubercles, and there is a faint streak of white on the tail. *Dorsal tentacles* long, linear, wrinkled, pellucid, slightly tinged with green and thickly spotted with white near the top: about one third down they are encircled by a dark brown band. They are set rather close together at the base, and held nearly parallel: the eyes are rather small and situated close behind them. *Oral tentacles* about half the length of the dorsal ones, linear, arising from the upper surface of the lips, whitish with an imperfect brown band in the centre; the tips obtuse. *Branchiæ* arranged in eight transverse rows on each side, the three anterior ones close together, the others wide apart. The first row contains three small papillæ, most of the others have four papillæ each; those next the back longest, the outside ones very small. They are large, linear, slightly elliptical, and pointed above, standing up in a fan-like manner across the back: the central vessel is of a pale warm green, and nearly fills the sheath, which has a brown ring at the base and some pale freckles of the same colour apparently arranged in three belts above; the surface is likewise freckled with pale opake yellow spots, crowded near the top, where they form an indistinct ring: the tip is transparent. The gastric vessel is seen of a green colour through the skin on the centre of the back, with branches along the base of the papillæ. *Foot* very narrow, linear, the anterior margin slightly bilobed and a little widened and rounded at the sides: the tail terminates a short way behind the branchiæ.

So nearly do the species of this genus approach each other, that it is often difficult to point out the characters by which they may be distinguished. A difficulty of this kind occurs to us in comparing this lovely little species with the *Eolis (Montagua) viridis* of Forbes. There are several small differences in colour and markings, but perhaps the best distinguishing characters are to be found in the close approximation of the first three rows of papillæ in *E. amœna,* and the brown markings on the back. The tubercular nature of the spots on this little animal is different from anything we have before observed in the genus. This species is also nearly allied to *E. Northumbrica,* from which, in addition to the characters already pointed out, it differs in the form of the papillæ and the greater length of the dorsal tentacles.

Two specimens were taken by the dredge in Torbay, in the months of May and June, 1845. They were rather inactive, and made very little progress when swimming. The foot seems well adapted for grasping corallines, but scarcely fitted for moving on a plain surface.

The spawn is white and deposited in a waved thread, forming a spiral of two or three turns.

Fig. 1, 2, 3. *Eolis amœna*, different views.
    4. Two of the papillæ more highly magnified.
    5. Spawn.
    6. A portion of the same more highly magnified.

Fam. 3, Plate 42.

PROCTONOTUS MUCRONIFERUS, Alder and Hancock.

P. ovatus, fulvidus, brunneo-marmoratus: branchiis ovatis, hyalinis, tuberculatis, in seriebus 12, utrinque ad marginem dorsi dispositis: papillis 4, magnis, frontalibus: tentaculis dorsalibus sub-tuberculatis.

Venilia mucronifera, Ald. and Hanc. in Ann. Nat. Hist. v. 13, p. 161, pl. 2.
Proctonotus mucroniferus, Ald. and Hanc. in Ann. Nat. Hist. v. 13, p. 407.
Hab. In shallow water, Malahide Bay, near Dublin.

Body nearly half an inch long, ovate, rather broad and depressed, subtruncated in front, and produced behind into a long pointed tail; the sides flattened or concave, projecting in a ridge above. The back is slightly rugose, of a pale yellowish brown colour, clouded and freckled with darker brown, and covered with minute opake white spots; the rest of the body is hyaline white, nearly colourless, having a few small brown spots on the head and sides, and more numerous opake white ones over the whole surface. *Head* with a semicircular veil, strongly notched in front, and bearing two short cylindrical tentacles at the sides. *Dorsal tentacles* of a purplish brown colour, with darker freckles, rather long, stout, and slightly conical, irregularly wrinkled and somewhat tuberculated; their bases nearly approximate, the points stand apart and are inclined forwards. Eyes rather large and placed as usual. *Branchiæ* ovate or inversely pear-shaped, produced into blunt and flattened apices, and having large, rather distant tubercular points over the whole surface. They have a crystalline appearance, being very transparent and nearly colourless, sprinkled with opake white spots. The central vessel is small and not half the length of the papillæ. It is yellowish, granular, and elliptical, tapering to a slender duct below. The branchiæ are set along the projecting ridge on each side of the back, in twelve transverse rows of three very close-set papillæ each; those next the back large and inflated, the exterior ones very small; two larger than the rest are placed posteriorly. These lateral rows are united anteriorly by four large, elliptical, tuberculated papillæ, apparently of a similar nature with the others, which pass round the head in front of the dorsal tentacles, and alternate with five smaller ones below. *Foot* rather broad and ovate, tapering to a fine point behind, transversely grooved and bilobed in front, but not produced into angles; it is pellucid, with a portion of the gastric system seen through, yellowish, and minutely spotted with opake white.

A single perfect specimen of this curious animal, and another much injured, were dredged up in Malahide Bay, in September, 1843, adhering to a sponge (*Halichondria panicea*) from rather shallow water. The first lived with us for two or three weeks. It turned sickly soon after it was caught and lost several of its larger papillæ: some of these were afterwards reproduced and grew very rapidly; but the papillæ on the front of the head, which were amongst those that had fallen off, never reappeared, so that we were prevented from satisfactorily ascertaining their functions; they seemed to want the central coloured vessel of the lateral papillæ, but perhaps might be considered equally with the others to perform the office of branchiæ. Our captive was not of active habits; it moved seldom and never quickly, and when it floated in the usual inverted position, which it did occasionally, it made no exertion whatever to assist its progress.

Fig. 1, 2, 3. *Proctonotus mucroniferus* in different positions.
4. A row of the papillæ more highly magnified.

Printed in May 2019
by Rotomail Italia S.p.A., Vignate (MI) - Italy